Flowers Coloring Book for Adults

Copyright © 2018 ColoKara
All rights reserved.

No part of this publication may be copied, reproduced in any format, by any means, electronic or otherwise, without prior consent from the copyright owner and publisher of this book.

Shutterstock authors: Credit images to - Thanakorn Hongphan, Patel B K, Voronin76, Todd Boland, Tiger4214, Pixy, PremiumStock, Chekmareva Irina, lisima, Africa Studio, Wacomka Alexey Vl B, Oleksandr Rybitskiy, Lumia Studio, Sunny Designs, MSNTY, Gringoann, Lisla, Sokolova_sv, Lisla, LuFei, Natali777, KostanPROFF, Anastasia Lembrik, MSNTY, Cat_arch_angel, Soyka, Ann.and.Pen, Nadezhda Molkentin, 7th Son Studio, GOLFX, Pawaris Pattano09, Olga Kleshchenko, Rednex, Pawaris Pattano09, Chekmareva, FLoric, Nadezhda Molkentin

Here's a special gift from ColoKara for purchasing this coloring book!

You can now freely download these coloring pages at any time and print them out as many times as you want!

Get your FREE coloring pages from this link: https://colokara.com/flowers-adult-green

HAPPY COLORING!

Made in United States
North Haven, CT
06 March 2022